**ROYAL ARMOURIES**

# CONTENTS

# HISTORY OF THE ROYAL ARMOURIES

**T**he Royal Armouries is Britain's oldest national museum, and one of the oldest museums in the world.

It began life as the main royal and national arsenal housed in the Tower of London. Indeed the Royal Armouries has occupied buildings within the Tower for making and storing arms, armour and military equipment for as long as the Tower itself has been in existence.

Although distinguished foreign visitors had been allowed to visit the Tower to inspect the Royal Armouries from the 15th century at least, at first they did so in the way a visiting statesman today might be taken to a military base in order to impress him with the power of the country. In the reign of Queen Elizabeth I less exalted foreign and domestic visitors were allowed to view the collections, which then consisted almost entirely of relatively recent arms and armour from the arsenal of King Henry VIII. To make room for the modern equipment required by a great Renaissance monarch Henry had cleared the Tower's stores of the collections of his medieval predecessors.

The Tower and its Armouries were not regularly opened to the paying public until King Charles II returned from exile in 1660. Visitors then came to see not only the Crown Jewels but also the 'Line of Kings', an exhibition of some of the grander armours, mounted on horses made by such sculptors as Grinling Gibbons, and representing the 'good' Kings of England, and the 'Spanish Armoury', containing weapons and instruments of torture said to have been taken from the 'Invincible Armada' of 1588. The Royal Armouries had become, in effect, what it has remained ever since, the national museum of arms and armour.

During the great age of Empire-building which followed, the collections grew steadily. Until its abolition in 1855, the Board of Ordnance, with its headquarters in the Tower, designed and tested prototypes, and organised the production

*The Line of Kings, a pen-and-wash drawing by Thomas Rowlandson, about 1800*

*Sir Henry Lee, Master of the Armouries, 1578–1610*

*Block and axe used for the execution of Simon, Lord Lovat, in 1747*

**"The armoury where all manner of arms are kept in readiness"**

KING CHRISTIAN IV
OF DENMARK

of huge quantities of regulation arms of many sorts for the British armed forces. Considerable quantities of this material remain in the collections today, and some can be seen on the walls of the Hall of Steel. Also, throughout this period trophies of all sorts continued to be sent to the Tower and displayed as proof of Britain's continuing military successes.

Early in the nineteenth century the nature and purpose of the museum began to change radically. Displays were gradually altered from exhibitions of curiosities to historically 'accurate' and logically organised displays designed to improve the visitor by illuminating the past. As part of this change items began to be added to the collection in new ways, by gift and purchase, and this increased rate of acquisition has continued to this day.

In this way the collection has developed enormously, the 'old Tower' material being joined in the last 150 years by the worldwide material which now makes the Royal Armouries one of the greatest collections of its type in the world.

As the museum's collections continued to expand the Tower became too small to house it all properly. In 1988 the Royal Armouries took a lease on Fort Nelson, a large 19th-century artillery fort near Portsmouth. This is now open to the public and displays the collection of artillery.

In 1990, after two years of preliminary research and deliberation, the decision was taken to establish a new Royal Armouries in the north of England in which to house the bulk of the collection of worldwide arms and armour, thus allowing the Royal Armouries in the Tower to concentrate upon the display and interpretation of those parts of the collection which directly relate to the Tower of London. The concept of the Royal Armouries in Leeds had been born.

*13-inch mortars at Fort Nelson*

*The Armouries as an arsenal: the former Cannon Room in the basement of the White Tower*

*The White Tower, the oldest part of the Tower of London, built about 1078–1100*

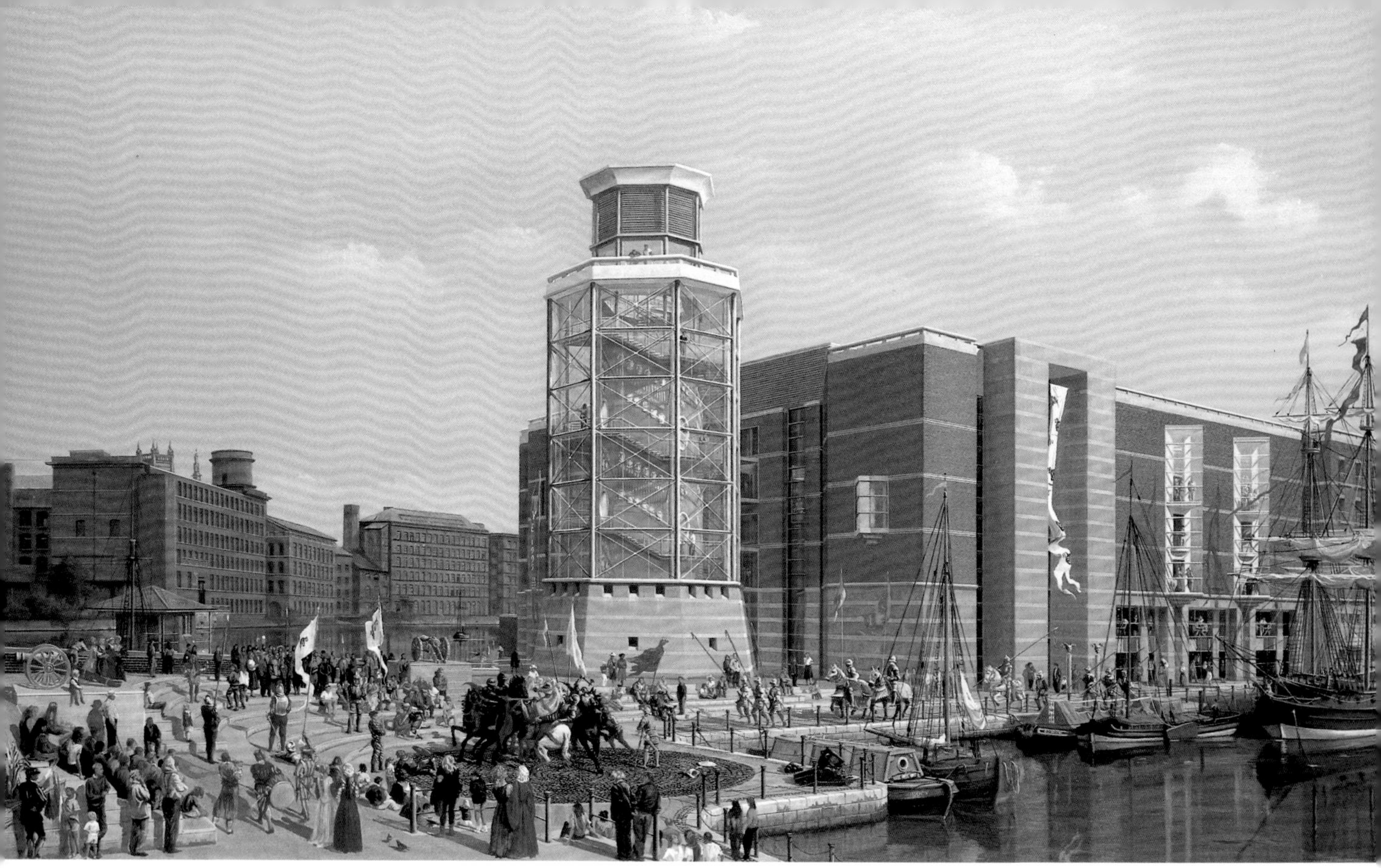

## DREAM TO REALITY

The new museum has been developed specifically to show the collections of the Royal Armouries in the best possible way. We began with the question 'How do we want to display our collections?', and the answer to that has dictated the sort of building which has been designed and built.

Colour, style, informality have been the watchwords for the design of the whole museum. One design team covering all

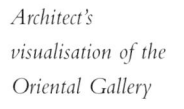

*Architect's visualisation of the Oriental Gallery*

4

disciplines has worked on the project from the beginning, each member gaining inspiration from the others and at the same time contributing to their work. It is the first time in this country that every aspect of a museum has been designed from scratch in this way. We hope you appreciate the result.

The Royal Armouries Museum has been built for the 21st century using the best of traditional museum design, and it has been developed quite consciously to show its collections in relation to the real world in which we live. The displays seek to make the historical stories relevant by bringing them up to the present day. The building reinforces this by allowing the modern world to permeate into the galleries in which those stories are told.

The building has, quite literally, been designed around the collections of the museum. The displays are intended to entertain and stimulate a desire to learn. This has meant developing a balance between standing and sitting, watching and doing, a balance which prevents aching feet and the yawn of boredom which all of us have experienced from time to time in museums.

Every gallery incorporates an appropriate mix of display and communication techniques. This mix ranges from the traditional museum display of objects in cases to the live demonstration of the use of weapons and armour. The objects are set in a context by themed interior design, by guides and interpreters, and by film and other audio-visual aids. There is an emphasis on 'hands-on' learning, so that by contact with real or replica objects, and by interactive computer programmes, the visiting public can gain a real feeling for and understanding of the collections. The intention has been to create a multi-layered experience to cater for the many different interests and interest levels of our visitors.

The use of violence by humankind for supremacy or survival, or its sublimation into sport or play always has been, and probably always will be, one of the main forces for historical change. This is the underlying theme of the new Royal Armouries. It is a fascinating and disturbing story of great importance to us and our children.

*View of the interior of the Street*

*Work in progress on the site, 1995 (Paul White Photography, Wakefield)*

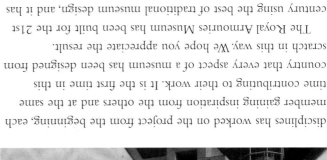

*Opposite: artist's impression of the finished museum by Carl Laubin, 1994*

*Architect's model of the museum complex*

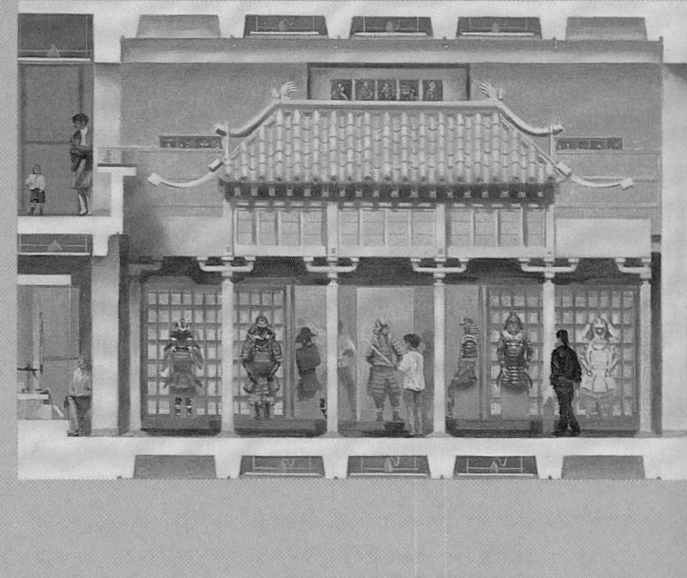

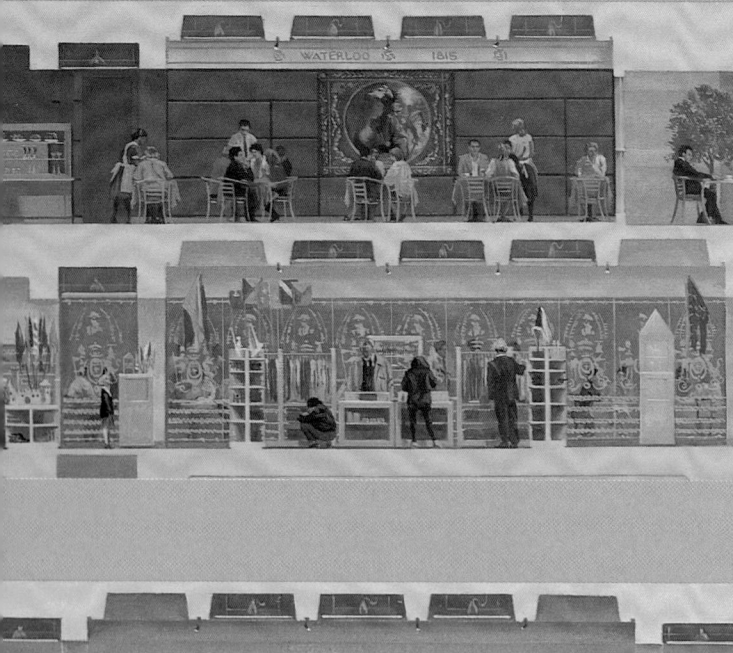

A MUSEUM FOR T

NO WAR OR BATTLE'S SOUND · WAS HEARD THE W

WATERLOO 1815

YAL ARMOURIES

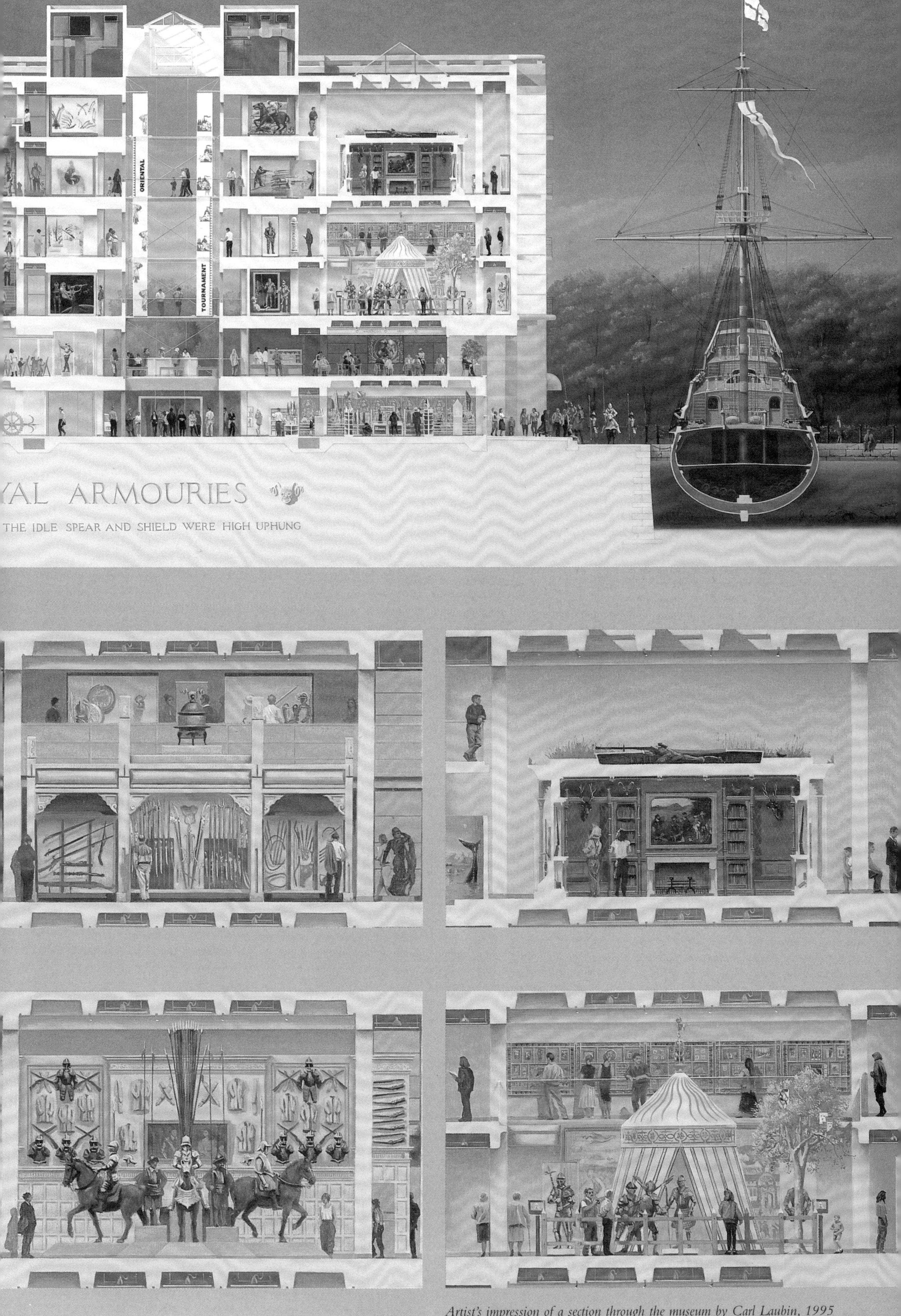

THE IDLE SPEAR AND SHIELD WERE HIGH UPHUNG

*Artist's impression of a section through the museum by Carl Laubin, 1995*

# WAR

>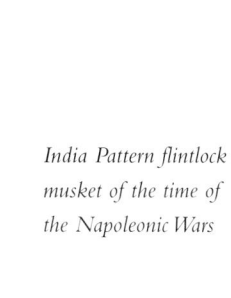
>
> **"Horrible war, amazing medley of the glorious and squalid, the pitiful and the sublime, if modern men of light and leading saw your face closer, simple folk would see it hardly ever"**
>
> **WINSTON SPENCER CHURCHILL**

**W**ar is one of those activities which most distinguishes the human species from other animals. Throughout the ages humankind has devoted much of its energy, ingenuity and power of social organisation to the systematic destruction of its own kind for reasons which may become incomprehensible with the passage of years but which were passionately felt at the time. As a result this century has seen humankind come within a button-push of utter destruction.

The face of battle seems to have changed enormously from the days of bloody hand-to-hand combat with which the gallery begins to the modern sophisticated, technology-based warfare witnessed recently in the Gulf War. This has largely been due to the enormous developments in Western science, technology and industry since the Renaissance five centuries ago. It is easy to be seduced by the clinical precision and distant impersonality of modern 'high-tech' war and to forget

*A still from Agincourt, one of the museum's 42 specially commissioned films*

*Opposite: armour for man and horse in the German 'Gothic' style, late 15th century*

*India Pattern flintlock musket of the time of the Napoleonic Wars*

that the majority of wars today are bloody, close, hand-to-hand affairs fought with relatively 'low-tech' weapons. For the inhabitants of the areas in which they are fought such warfare causes the world to be turned upside down just as it has from the beginning of human history. These wars have not really changed, even though the weapons have, and the act of warfare remains ever the same and essentially de-humanising. Yet, and here lies one of the fascinations of the study of war, in this apparently barren environment men and women through the ages have often displayed the most amazing courage and fortitude and the most sublime humanity and self-sacrifice. This gallery is their story.

To understand why generation after generation seems willing and often eager to go to war, the gallery begins by exploring the seductive myths, pomp and circumstance of the heroic soldier. It is an image fostered by generations of recruiting sergeants which brings a tear to the eye and a lump to the throat when a military parade passes by. But it is an image which does not stand up to the reality of war.

For the soldier, once enlisted, the reality of warfare has always been very different from the heroic gloss put on it by the recruiting sergeant. So in a 150-seat cinema visitors are brought face to face with the terrible realities of war, starting with the horrors of modern battle and ending with the bloody hand-to-hand combat with which the main gallery begins.

The main gallery tells the story of the weapons and defensive armour used to fight wars across the centuries by explaining the leap-frogging march of technology in offensive arms and defensive counter measures, and by chronicling the

*Scene from a German firework book, about 1440*

*The battle of Pavia, 1525, painted in Italy at about that time*

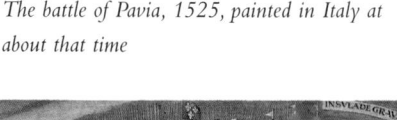

*The decisive moment of the battle of Pavia, 1525. The French cavalry is defeated by Imperial arquebusiers*

*Great helm, English, about 1370*

changing face of battle for those who had to endure it.

From Classical times and throughout the Middle Ages armour proved a reasonable defence against the cut of sword, the thrust of spear, and the impact of arrow. As weapons improved in effectiveness those who could afford it replaced mail with plate armour. Full plate armours were made by highly skilled armourers who literally tailored the armour to fit the wearer, allowing him complete freedom of movement. On his own a fully armoured knight could easily mount and dismount his horse. Suggestions of the need for cranes and of the inability of knights to get up if they fell off we owe to Mark Twain and Hollywood not to the reality of warfare in the Middle Ages. With the increasing use and effectiveness of firearms in the 16th and 17th centuries the use of armour gradually declined, though it was still much used in the English Civil Wars (1641–53) and never entirely died out. The use of armour has gradually revived in this century (for instance, for use by machine gunners on land in the First World War and in the air in the Second) and today, once more, all British soldiers are armoured for combat.

From humble beginnings as little more than a noise and smoke machine on the 14th-century battlefield the gun had developed by the time of the American Civil War into a weapon which, when used in sufficient numbers, could scythe down a field of corn, and with the corn any poor soldier who was trying to cross it. To begin with it proved a slow developer. It was not until the 16th century that it really began to be a

*Equipment for a harquebusier (light cavalryman), English, about 1650, from the armoury at Littlecote House*

battle winner. In 1525, at Pavia in northern Italy King Francis I of France thought he was about to win the battle with a good, traditional cavalry charge, only to find that his armoured knights were brought down by the German infantry armed with matchlock guns. In a puff of smoke victory turned to defeat and it was obvious that the days of chivalry were at an end.

From then until the middle of the 19th century the soldier's gun remained a simple affair which in essence changed little. But then came a period of revolution in design brought about by the application of science and technology to military affairs. In under 50 years the infantryman's musket was transformed from a slow, inaccurate short-range muzzle-loading single-shot gun to a rapid-shooting, accurate long-range breechloading repeater. The last years of the century also saw the arrival of the fearful weapon which in World War I was to change the face of warfare – the automatic machine gun. Invented by Hiram Maxim it was first used in the colonial wars of the

*A Gatling machine gun, British, made in 1873*

*A British machine gunner of the First World War*

1890s and its immediate supremacy over other weapons led Hilaire Belloc to write: 'Whatever happens we have got the Maxim Gun and they have not'. It was to be Europe's tragedy in World War I that both sides had the machine gun and it and massed artillery were chiefly responsible for the appalling casualties of that war.

*The seductive myths of war continue. Now it is war among the stars that seems clean and heroic*

# TOURNAMENT

**"Trulie, this action was mervailouslie magnificent, & appeared a sight exceeding glorious"**

FROM A DESCRIPTION
OF A TORCHLIGHT
TOURNEY HELD IN
WHITEHALL IN 1572
IN THE PRESENCE OF
QUEEN ELIZABETH I

**T**he story of the tournament is the story of how a form of early medieval practice for war, which was generally frowned on by monarchs as a danger to public order, developed into perhaps the greatest of the Renaissance courtly entertainments used by monarchs to show off their power, wealth, and chivalric ideals. From uncontrolled beginnings the tournament evolved into a highly regulated and sophisticated sport with as many different varieties as football. From the 15th century onwards special weapons and very specialised armours were required for the various types of tournament, and the Royal Armouries contains a fine collection of these. These are exhibited and interpreted in this gallery which charts the development of the three major types of tournament fighting – the tourney, the joust and foot combat.

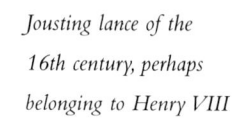

*Jousting lance of the 16th century, perhaps belonging to Henry VIII*

*Foot combat with pollaxes, from a French manuscript, about 1475*

*Opposite: tournament armour from the court of Maximilian I, about 1500*

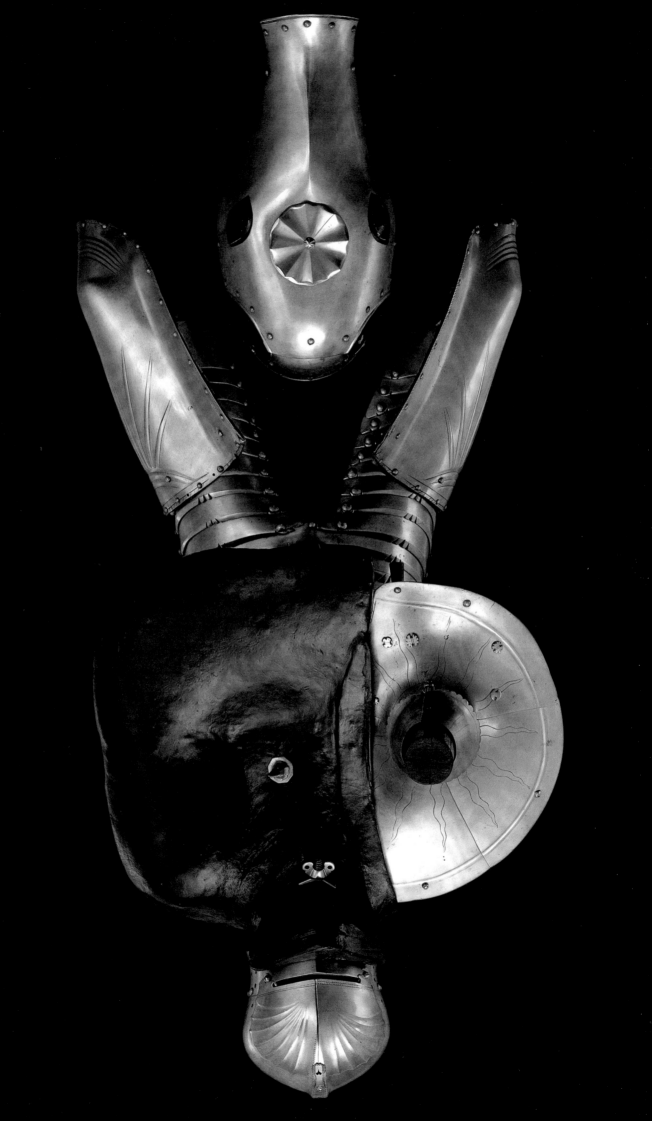

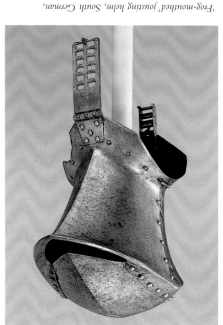

*The Field of the Cloth of Gold (The Royal Collection © Her Majesty the Queen)*

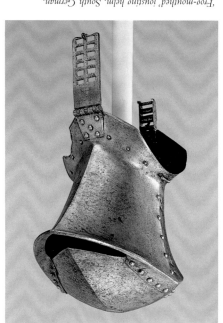

*'Frog-mouthed' jousting helm, South German, about 1480*

The tourney is the earliest form of tournament combat. It was developed in the late 11th century, at a time when in battle knights were using a heavy lance held tightly under the right arm. This enabled the rider to put his full weight behind his lance and to unhorse his opponent without losing his own weapon. The tactic was to charge, break the opposing ranks and then close with the sword. The tourney provided the ideal opportunity for the individual to practice his horsemanship and his management of an unwieldy weapon as part of a team in circumstances that approximated to those of real warfare. The tourney was therefore fought in an open field with two teams of mounted combatants who charged with lances and then fought with swords. Later many variants developed, as well as special armour. This was based on that for war, but with heavy reinforces on the left side for the initial charge with lances; these were often removed for the second charge with swords. A distinctive locking gauntlet was worn for the sword attack.

The joust developed the knight's skill in controlled horsemanship and lance work. It was a contest between two single, mounted combatants who charged each other with lances an agreed number of times. Each knight would fight for himself in a restricted space marked out by fences. Commonly, from the middle of the 13th century distinctions were made between jousts *à plaisance* (later called jousts of peace) and jousts *à outrance* (later called jousts of war). The distinguishing feature was the rebated (or blunted) lances used in the joust of peace while the joust of war used sharp lances. The so-called frog-mouthed helm was the first identifiable piece of armour

*Screen from one of the computer interactives in the Tournament Gallery*

**HERALDRY**

This ... is correct. Any design or charge can be used though not many would choose a hamster. Lions and eagles were popular during the Middle Ages as symbols of aggressive courage.

Click to Continue

*The tonlet armour of Henry VIII, English, 1520*

designed exclusively for the joust. It curved up and out at the front to protect the eyes and deflect a jarring blow. Also, as part of a growing emphasis on safety, a tilt barrier was introduced, which separated the horses. The chest and head were reinforced and locked into a forward-facing position.

The foot combat was a contest between two single combatants who fought each other on foot with either axes,

*Foot combat in the 15th century; a scene from the introductory film to the gallery*

two-handed swords, spears or other agreed weapons. Usually a set number of blows was delivered by each. The classic foot-combat armour protected the whole body, with both sides equally protected. The armours were therefore symmetrical, with either a cod-piece or a steel skirt, called a tonlet. The 'barriers' was a form of foot combat in which the combatants were separated from each other by a barrier.

The gallery has as its central feature an area in which the art of foot combat is demonstrated by interpreters dressed in replica armour of about 1500 who literally pollaxe each other. Here too you are allowed to handle some pieces and replicas and learn at first-hand with the help of an interpreter what it was like to use and wear them. In addition there are major displays on the flowering of the tournament in Tudor times, including the greatest tournament event of all, the Field of Cloth of Gold, and the creation of the cult of Elizabeth I as the Virgin Queen, which was largely the achievement of a previous Master of the Armouries, Sir Henry Lee, through his Accession Day Tilts.

# ORIENTAL

**"Military action is important to the nation – it is the ground of death and life, the path of survival and destruction, so it is important to examine it"**

SUNZI
*THE ART OF WAR*
ABOUT 4TH CENTURY BC

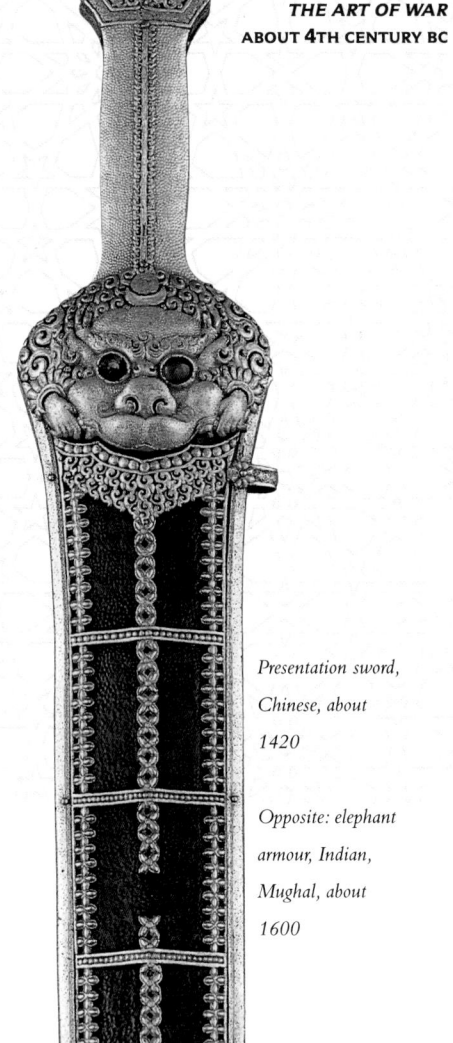

*Presentation sword, Chinese, about 1420*

*Opposite: elephant armour, Indian, Mughal, about 1600*

**A**rms and armour have been made and used around the world for war, sport, military practice and self defence. This gallery concentrates upon the great civilisations of Asia, and its purpose is to show how arms and armour can provide a key to understanding Asian history, a subject which until recently in Europe has been largely ignored.

The Royal Armouries acquired its first Asian piece in the 17th century – one of two armours presented as diplomatic gifts to King James I by the Shogun of Japan, Tokugawa Ieyasu. This was first displayed in the Tower in 1660. More recently it has been joined by the other and now both, the first two Japanese armours seen in England, are on display.

The cultures of Asia are far more diverse than those of Europe, and the gallery is divided into a number of distinct zones – central Asia, Islam, the Indian sub-continent, China, Japan, and South-East Asia. But there is one military theme that unites all these diverse cultures, and has dominated the way in which war was waged in most of them until the 19th century: the use of the mounted archer.

The cavalry armies of the empire-building cultures of Central Asia, the Scythians, the Huns, the lesser-known Xianbei, the Turks and, most famous of all, the Mongols, were all composed of horse archers. Their combination of mobility and fire-power was awesome. It enabled them to carve out empires in the rich lands of the south and west, and caused their neighbours to imitate their military systems – they simply could not beat them, so they joined them.

After the formation of the great Islamic empire, its political unity gradually weakened, though the religious fervour that had formed it remained, and the Muslim religion continued gradually to expand. Converts were welcomed, none more than the mercenary Turks from the north, who infiltrated at first as slave-soldiers and later as independent kingdoms. The most successful of these were the Ottomans, who in the 15th century put an end to the Byzantines, rulers of the old eastern Roman empire, capturing the capital, Constantinople, in 1453 and renaming it Istanbul. By the 17th century they were the principal Islamic power in the West, controlling North Africa, the Middle East and the Balkans.

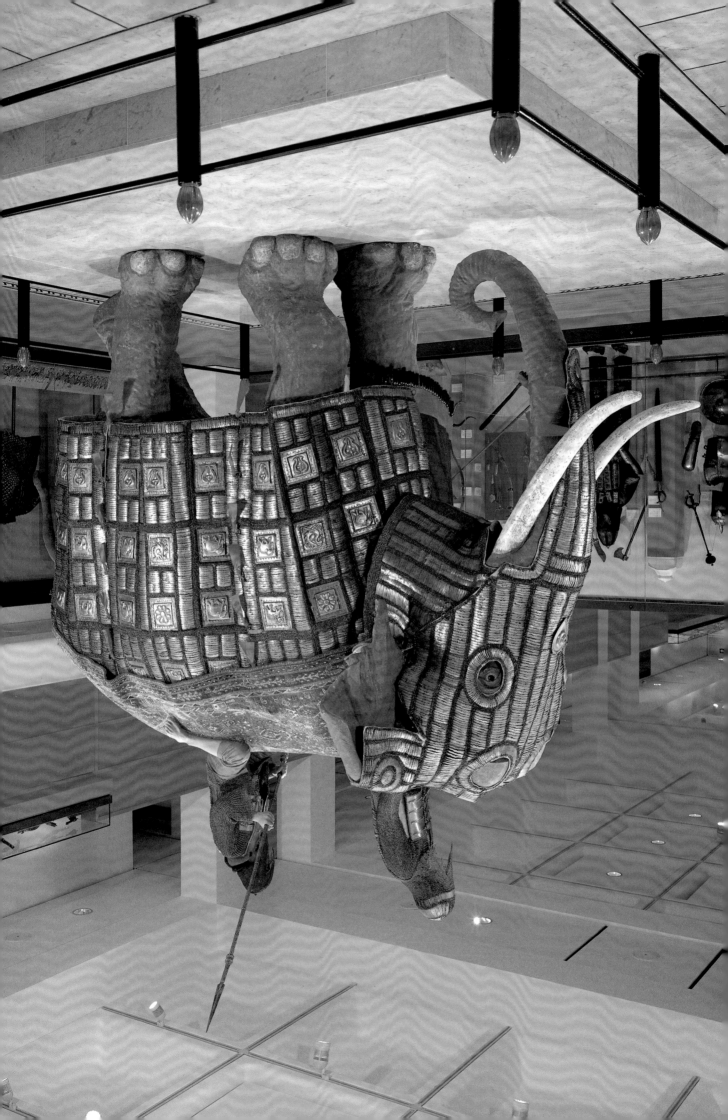

Arms and armour from India form the largest part of the Royal Armouries Asian collection. In the 19th century British interests in the sub-continent, through the East India Company, were substantial, and allied with this military and commercial activity came scholarship and a desire to educate the populace back in England about the wonders of India. (We use 'India' in the museum as a generic term to cover the current political divisions of Bangladesh, India and Pakistan, without any intention to insult or slight. In the same way we use the old term 'Persia' to refer to present-day Iran and Iraq, nations which are now independent but for much of the period covered by the museum were united.)

Quoit turban (dastar bungga), Indian, Lahore, 18th century

The line of equestrian figures in the Oriental gallery

From India comes the largest armour in the collection, the only elephant armour in captivity. It was probably made in one of the arsenals of the Mughals in north India in the late 16th or early 17th century, and was acquired in India in 1801 by the second Lord Clive, Earl of Powis. The armour is of mail and plate construction, and in its present state weighs 142 kg: six of

Lamellar armour for man and horse, Central Asian, 15th–19th century. Horse armour on loan from the Victoria & Albert Museum

China has been responsible for a large number of inventions which have influenced world history. Pre-eminent among these is the invention of gunpowder; its first recorded military use was in 919 AD and the first published recipe for true gunpowder appeared in 1044. Another Chinese invention which was very influential in the West was the crossbow, first used in the 4th century BC.

It is perhaps in Japan, however, that arms and armour have the most prestigious history. The Japanese sword, perhaps the most excellent cutting weapon ever devised and certainly the best studied and preserved, is often described as the 'soul of the samurai'. The Shinto religion, with its belief in the survival of the spirit in the artefacts associated with its life, has encouraged the preservation of the armour and weapons of great warriors, and more medieval arms and armour are preserved today in Japan than in the rest of the world put together. *Yabusame*, the art of horse archery, is still practised at the Nikko Toshogu Shrine, with which the Royal Armouries has been twinned since 1991. The Shrine was built in 1623 as the resting place for the spirit of the Shogun who reunited Japan after centuries of civil war, Tokugawa Ieyasu: none other than the ruler who presented King James I with two armours in 1613, displayed here in the Royal Armouries today, one of which is illustrated on the right.

its eight pieces survive (two of the three panels for the right side are missing). Originally it would have weighed 170 kg.

*Yabusame at the Nikko Toshogu shrine*

# HUNTING

HUNTING

**"There is a passion for hunting, something deeply implanted in the human breast"**

**CHARLES DICKENS**

*American powder-horn of about 1770. Made of cow-horn it is engraved with a map of the Mohawk and Hudson rivers*

*Opposite:*
*'An Ugly Customer'*
*– a recreation of a tiger hunt in India in the mid 19th century*

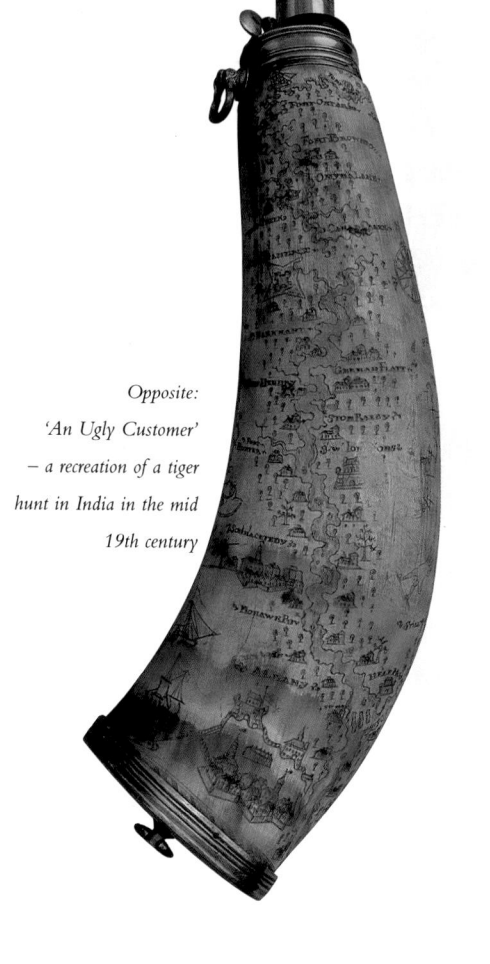

**T**he human species began as hunter-gatherers, and ever since men and women around the world have hunted for subsistence, for profit or for sport. Hunting has had a profound effect upon landscapes and ecologies, and has both preserved and destroyed whole species of animals and birds. Hunting has bestowed on us a great tradition of legend and literature, and a rich heritage of music. It has also given us many of the breeds of horses and dogs which we own and love today. All types and conditions of mankind have hunted and do hunt.

*The charge, a print from* Hog-hunting in Lower Bengal *by Percy Carpenter, 1861*

There are many images of the hunter, from the self-reliant Leather Stocking living in harmony with nature in his backwoods to the callous big-game hunter on his relentless and insatiable pursuit of trophies. To many today hunting is repulsive and indefensible, to many it is right and natural, and to some it is still essential for survival. This is the story, both good and bad, which is told in this gallery.

About 10,000 years ago man first began to domesticate animals and plant crops. From settled farming communities producing an excess of food came the ability to specialise, to learn other craft skills, to live in towns, to become civilised. The change to settled agriculture and the development of civilisation led to increased populations which ever since have threatened animals by destroying their habitats.

Despite the change to a settled agricultural existence with its much reduced need to catch food to survive, man's love of hunting continued. It was in the blood.

In Medieval Europe rich and powerful landowners jealously guarded their rights to hunt. Great estates and forests were

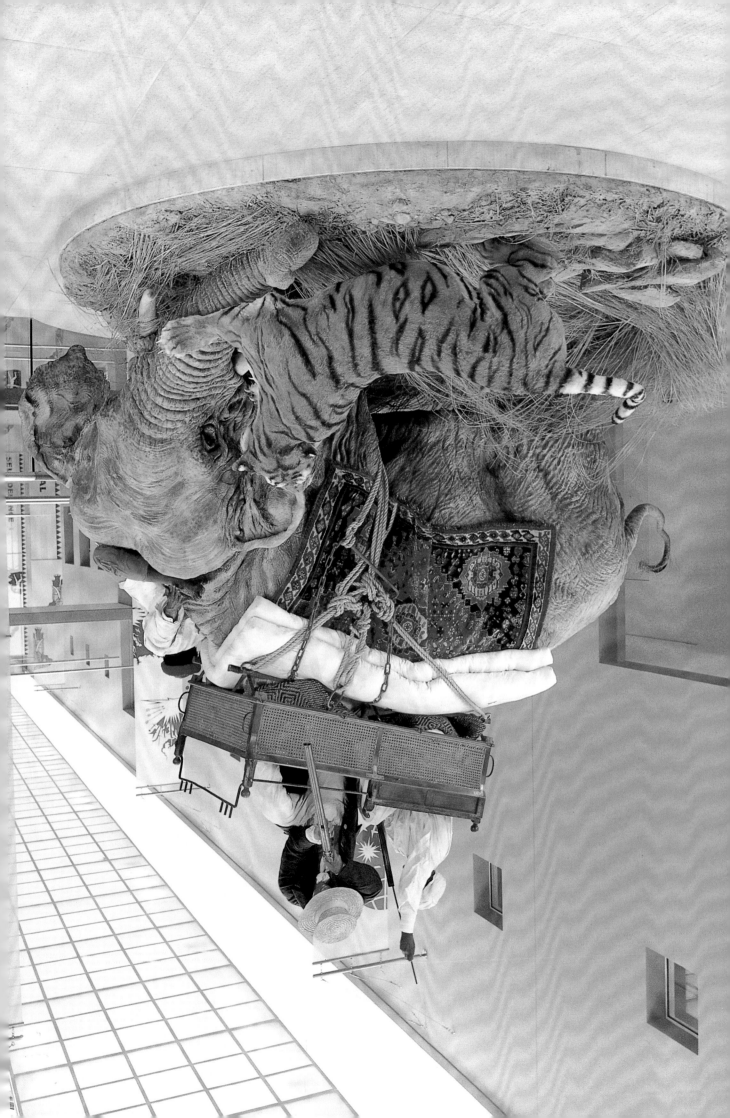

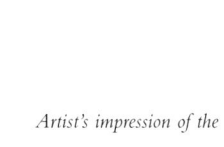

carefully maintained where animals were hunted as a social and fashionable pastime. The hunt was part of the sophisticated chivalric culture of the Middle Ages, as much a part of court life as banquets, music, dancing, and courtly love.

In Europe in the later Middle Ages the traditional hunting weapons of bow and crossbow began to be challenged by fast-improving firearms – more and more effective weapons were developed allowing animals to be hunted from greater ranges with less risk and more certainty of a kill.

New and improved weapons led to the development of new types of hunting. In the 17th century light and accurate firearms were developed which allowed birds to be shot in flight. By the 19th century shooting weekends on large estates were great social occasions.

The shooting party became fashionable and many sportsmen and women spent much of their spare time shooting large numbers of birds. On his estate near Harrogate in 1888 Lord Walsingham shot more than 1,000 grouse in one day. But the numbers of game birds in Britain did not decline as a result; rather, they increased. Estates were carefully managed to provide good sport, birds were specially bred, nurtured and protected from natural predators as they are today.

Elsewhere in the world game management has not always been so successful. As Europe came to dominate the world and its technology reached into every corner of the planet, European methods of hunting were added to the often highly efficient hunting methods of native tribes and cultures. In the

*Hunting spear, French or possibly Italian, about 1600*

*Artist's impression of the Gun Room*

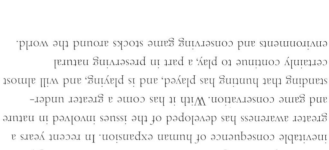

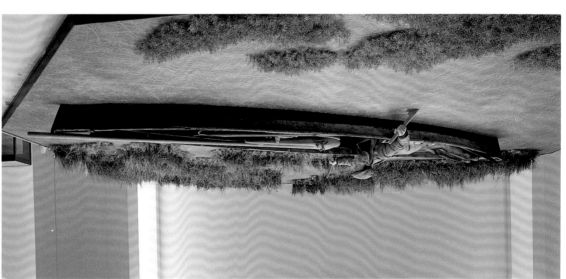

*Flintlock gun and powder flask from a set of hunting guns and accessories made for Empress Elizabeth of Russia in the Imperial arms factory at Tula in 1752*

*Tableau of the Essex wildfowler Walter Linnett on the River Blackwater, about 1920*

19th century in the Far East, India and Africa the numbers of big game reduced as a result. In North America the vast numbers of buffalo that roamed the plains were reduced with terrifying speed to a few small herds. In 1868 a railroad traveller recorded seeing a herd of buffalo 120 miles long. Only 16 years later not a single head of 100 head survived. The buffalo had become the first species to be driven to the brink of extinction by the development around it of an entire industry. In the 20th century the same fate has befallen the whale, and ivory-bearing animals, like the elephant, and the rhinoceros.

The pressure of population growth puts the existence of entire habitats and species at risk as a direct and seemingly inevitable consequence of human expansion. In recent years a greater awareness has developed of the issues involved in nature and game conservation. With it has come a greater understanding that hunting has played, and is playing, and will almost certainly continue to play, a part in preserving natural environments and conserving game stocks around the world.

## SELF-DEFENCE

**F**rom the caveman with his club, people have armed themselves against the threat of violence from others.

When this gallery's story begins, in Europe as it emerges from the Dark Ages, people had to protect themselves from outlaws and invaders as well as wild animals. But they were also quick to use weapons in anger or to settle quarrels. Gradually, legal systems took shape to which people increasingly turned and local peace-keepers were appointed.

Less successful were attempts to control the carrying of weapons. In England, even after a century of strong Tudor rule, people often went armed with staff, dagger or sword and the government tried in vain to check the spread of new, more dangerous weapons, such as the rapier and the handgun.

Increasing national prosperity meant more opportunities for the criminal, such as the smuggler and the highwayman. As England entered its Industrial Revolution old towns and villages were transformed and new ones suddenly created. Law and order traditionally based on small, stable communities seemed about to collapse.

One response to this danger to power and property was the new police, backed up by the new prisons. However, law-abiding citizens were not ready to put all their trust in these new protectors and so the ordinary citizen continued to keep

*A fencing exhibition, French, about 1820*

> "The right of each to carry arms – and these the best and the sharpest – for his own protection is a right of nature indelible and irrepressible"
>
> **JAMES PATERSON, 1877**

*Colt 'Peacemaker', American, about 1891*

weapons, which were now more easily and cheaply available as mass production turned out vast quantities of compact handguns.

Meanwhile, those who travelled abroad, beyond the protection of their own or any government, armed themselves as they saw fit against the threat of hostile local people or wild animals.

In 20th-century Britain the uncontrolled possession of weapons in private hands has come to be seen as a threat rather than a protection to the ordinary citizen. The police meanwhile have responded to new challenges with new weapons and equipment, at odds with the traditional image of the unarmed constable on the beat.

Even so, the fear of violence continues or is even growing, and there is now public recognition of what was until recently hidden violence – against women and children.

In the opening up of the American West that in Europe extended over centuries – the taming of violence and the establishment of law and order – was compressed into decades. Into this turmoil poured vast quantities of mass-produced firearms from the industrialising Eastern states. In the West, the gun helped to shape a new society, symbolising independence and freedom on the frontier, but also leaving a legacy of violence.

Weapons have been kept and carried in ordinary life for other reasons besides self-defence and law enforcement: for instance, for sports such as target shooting and fencing; for duelling in defence of aristocratic honour; as fashion accessories; or to symbolise authority and status and commemorate courage and service.

*The crossbow shooting gallery*

*Sword made for presentation by the City of London to Admiral Lord Collingwood, Nelson's second-in-command at Trafalgar. English, 1805*

*A scene from one of the gallery's films*

# NEWSROOM

> **"Acts of violence – whether on a large or small scale, the bitter paradox: the meaningfulness of death and the meaninglessness of killing"**
>
> DAG HAMMARSKJOLD

**A**ll the themes of the museum are brought completely up to date in the Newsroom where visitors can see the museum's own newscast of relevant events around the world, and can occasionally witness events as they happen. Modern news-gathering techniques and technology will be used to ensure the immediacy of the multimedia presentations which centre on the large 15-screen vidiwall. Occasionally these will be enhanced by special sessions in which specially trained interpreters will discuss and explain to visitors current events relating to the subjects of the museum. The aim is to ensure that every visitor understands the relevance of the museum to the world we live in today.

*Photographs: Steve Eason / Reuter / Hulton Deutsch*

# HALL OF STEEL

The Yorkshire Electricity Hall of Steel is the architectural centrepiece of the museum. It is both a major circulation spine and a magnificent display space.

It rises over 100 feet and consists of a hollow inner octagon, on both sides of which, above the eight slender supporting pillars which rise to first floor height, are massed wall displays of over 3000 pieces of arms and armour, a spiralling staircase and an outer octagon, made entirely of glass from the first floor upwards. From the city, especially at night when it is illuminated, the Hall stands as a beacon showing what the museum contains to all who pass. From inside it gives visitors who climb its stairs fine views over the city centre.

The idea of displaying battle trophies on the walls is an old one. Alexander the Great found arms and armour, allegedly used in the Trojan Wars, displayed in the temple of Athena at Ilium. In England arms and armour were displayed decoratively on the walls of houses from the time of Elizabeth I. But massed displays of many pieces first appeared on the walls of royal guard chambers shortly after the Restoration in 1660. The largest and most impressive displays ever were created in 1696 by John Harris to embellish the Small Armoury in the Tower of London. These used thousands of weapons and weapon parts to create tableaux including the backbones of a whale, a giant organ and a seven-headed monster.

The displays in the Yorkshire Electricity Hall of Steel consist mainly of 17th-century armour and 19th-century military equipment. They represent the largest mass display assembled since the 19th century. The designs were inspired by the formal compositions of the 17th century. The items displayed on the outside of the inner octagon can be viewed closely as you go up or down the stairs, and those on the inside can either be seen in all their massed impressiveness from the ground floor, or in detail through the porthole windows as you climb the stairs.

Because such wall displays were a feature of the historic displays in the Tower of London the Hall of Steel is a real link between the future and the past of the Royal Armouries. Whether gleaming in the midday sun, glinting in a rainstorm, or glowing at night, it reminds both us and the city of our ancient origins and our future commitment to our new home.

"No war, or battle's sound
Was heard the world
    around;
The idle spear and shield
    were high uphung"

**JOHN MILTON**

*Photo: Press Agency, York*

# CRAFT COURT

The new Royal Armouries is more than just a museum in a building. The objects in the collection were largely made for use out of doors, and therefore external demonstration and event areas have been provided to enable the collection to be properly put in context, explained and demonstrated.

Next to the main building is the Craft Court where the visitor can see a selection of craftspeople working at their trades in traditional workshops. These will include: an armourer working with a traditional charcoal forge, making armours as they were made in the Middle Ages; a gunmaker, making and repairing modern and antique guns, and combining the metal and woodworking trades which have traditionally formed the art of the gunmaker; and a leather-worker, making boots, saddles, and the buff coats that were used by so many soldiers for defence during the Civil Wars of the mid 17th century.

These craftspeople have been carefully selected by the Royal Armouries, and are among the very best in the country. They work largely with modern tools, wearing modern dress, but using traditional techniques. Visitors can watch them work at their craft, but it is important to realise that these people are here working for their living. They are not paid by the museum; they make their money by working at their craft.

*Chris Dobson, armourer in residence at the Royal Armouries, working at his forge*

*Venus at the Forge of Vulcan: an allegory of fire, by Jan Breughel and H van Balen, Flemish School, about 1620*

# TILTYARD & MENAGERIE

## Tiltyard

Running along the riverside for 150 metres from the museum, with stands on the landward side, is the Tiltyard for demonstrations of jousting and horsed combat of all types, skill at arms, military drill and manoeuvre, and hunting and animal-handling techniques.

During the summer months there are major performances each day and you may find that you have to take your place early to see one as spaces are limited. During the winter months or in the summer in adverse weather conditions the work of training the animals goes on and you may come and watch these training sessions for as long as you like.

Our intention in the Tiltyard is to give you an impression of how our collections were used which would be impossible inside the museum. What you see is performed largely using replica weapons and armour which the museum is constantly acquiring and improving at very considerable expense.

## Menagerie Court

Between the Tiltyard and the Craft Court is the Menagerie Court where the birds, dogs and horses are housed while working. Here you can see them at close hand and meet and talk to their trainers and assistants.

The Menagerie Court is named after the famous royal menagerie, or zoo, which was housed in the Tower from 1235 to 1831 when its transfer to Regents Park formed the foundation collection of London Zoo.

The horses are our own, specially selected to represent as accurately as we can a type of medieval war horse. You may be surprised at their size, but generally the medieval war horse was not particularly tall by present-day standards. But they were, like ours, strong, short-backed and with thick, powerful necks. Four of our horses – Fleur, Berwick, Messenger and Gauntlet – are Lithuanian draft horses, specially imported for us. They have all been rigorously trained to ensure that they are comfortable with swords and spears clashing next to their eyes and with the loud reports of firearms shot from just behind their heads.

Gauntlet is named after the armorial device of the Master of the Armouries and is the horse he rides on special occasions. The hawks, falcons and dogs are provided and worked for us by the Lakeland Bird of Prey Centre, Lowther Castle, near Penrith.

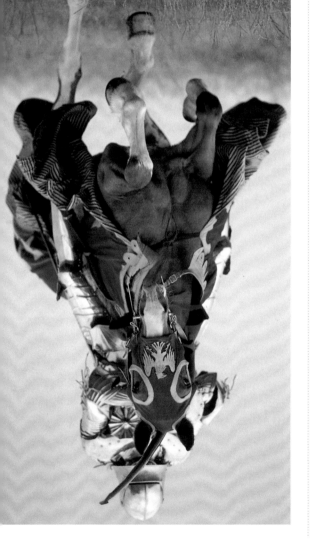

*Top: one of the museum's demonstrators at full tilt; below: James Buttle, owner of the Lakeland Birds of Prey Centre, working a peregrine falcon*

# FOOD AND SHOPPING

## The shop

Browse among our unique merchandise celebrating the collections, displays and themes of the museum. Original bespoke curiosities and mementoes make interesting souvenirs of your visit or unusual gifts for family and friends.

## The Nelson – ground floor

Have a meal, a snack, afternoon tea or just a glass of wine by the waterfront. Choose from a traditional but contemporary English menu.

## The Wellington – ground floor mezzanine

Overlooking Clarence Dock, this is our fast, self-service café, where freshly made sandwiches are our speciality. Those with a heartier appetite can try our 'meal in a bowl'. This is the perfect place for families.

## Coffee bars – first and second floors

Just the place to rest, relax and reflect on what you have seen so far. Treat yourself to freshly brewed coffee and one of our traditional Yorkshire cakes. Refreshed, continue on your way to explore another gallery.

## Other museum facilities

In addition to our extensive permanent displays, events are also often staged in the Bury Theatre, Tower Room and Royal Armouries Hall.

A fully equipped Education Department serves the needs of groups of all ages and includes a collection of objects that may be handled. Disabled visitors are welcomed: the museum has been developed with the disabled in mind and a Disability Access leaflet is available free on request.

Academic services include a world-class library, study collections and photographic archive, specialist conservation and analytical services, and a team of expert curators.

**"The victualling of the troops comes first as it must be done almost daily and for each individual"**

CARL VON CLAUSEWITZ, 1832

The Royal Armouries thanks its current sponsors for their encouragement and support:

 JVC

 PINSENT · CURTIS

 BT
*Community Programme*

HALIFAX

Yorkshire
Electricity